COMPETITIVE BIOLOGY 3

INTRODUCTION

This objective biology series provides a basic and challenging problem of biology from particular topics. It can be used to brush up ones basics and checking up the preparation level of particular topic. It is equally helpful to the traditional classes as well as competitions. It can be also taken as a revision material for any competition which includes the test of basic biology. If you want to grasp the subject before practicing these multiple choice questions, you can go through the website http://www.ncert.nic.in/ncerts/textbook/textbook.htm and down load the free copy of science books and after having command on the topic practice it. For revision purpose, important points are given at the starting of each topic.

If you have any query or suggestion about this series you can send your suggestion at uk2594@gmail.com.

CONTENTS

9. OUR ENVIRONMENT

SOME IMPORTANT POINTS

➢ Biodegradable substances can be broken down by the action of bacteria's.

➢ The non biodegradable substance present in environment for a long time.

➢ The ecosystem forms by the all interacting organisms in an area.

➢ An ecosystem includes both biotic and abiotic components.

➢ The organisms which make our food by inorganic substances are called produces.

➢ The organisms which are dependent on producers for their food are called consumers.

➢ The organisms are which break down the food remains of organisms are called decomposes.

➢ The levels of a food chain are called tropic levels.

➢ The food chain consists series of organisms feeding on one another with various tropic levels.

➢ The organisms which are the at first tropic levels are produces.

➢ The organisms which are at second tropic level are primary consumers.

➢ The secondary consumers form the third and tertiary consumers from fourth tropic level.

➢ The produces capture about 1% of sunlight energy to make its food.

➢ The 10% of energy taken as the average value at each step from producers to the next levels of consumers.

➢ The transfers of harmful chemicals at each tropic level is called biological magnification.

➢ When the three atoms of oxygen combined they form a molecule of ozone O_3.

➢ The ozone layer is not act as an protective blanket and protect us from UV radiation from the sun.

➢ The increasing in uses of CFCs from refrigerator resulted in ozone layer.

9. OUR ENVIRONMENT

1. A natural phenomenon that becomes harmful due to pollution is?

    a. Global warming   b.Ecological balance   c.Green house effect

2. The pollutant responsible for ozone holes is ?

    a.CO_2 b.SO_2 c.CO   d.CFC

3. One of the best solutions to get rid of non biodegradable wastes is?

    a.Burning   b.Dumping   c.Burying   d.Recycling

4. Animal dung is _____ waste.

    a. Biodegradable   b. Hazardous   c. None biodegradable   d. Toxic

5. Which of the following is biodegradable?

    a. Iron   b. Plastic   c. Lether belts   d. Silver

6. Which of the following is non biodegradable ?

    a.Animal bones   b.Nylon   c.Tea leaves   d.Wool

7. Name one non biodegradable waste which may pollute the earth to dangerous levels of toxicity ,if not handled properly ?

    a.DDT   b.CFC   c.PAN   d.Radioactive substances

8. In lake polluted with plastisides ,which one of the following will contain the maximum amount of pesticides ?

    a. Small fish   b. Big fish   c. Water birds   d. Microscopic animals

9. The major pollutant from automobile exhaust is ?

    a. NO   b.CO   c.SO_2 d. Soot

10. The green house gasses,otherwise called radioactivitely active gases includes:

    a. Carbon   b.CH_4 c.N_2O d. All of these

11. Algal bloom results in ?

    a. Global Warming   b. Salination

    c.Eutrophication   d.Biomagnification

12. A high biological oxygen demand indicates that ?

    a.Water is pure   b.High level of microbial pollution   c.Absence of microbial

13. The effects of radioactive pollutants depends upon ?

    a.Rate of diffusion   b.Enegry releasing capacity

    c.Rate of deposition of the contaminant   d.All of these

14. The range of normal human hearing is in the range of ?

    a.10Hz to 80Hz   b.50Hz to 80Hz   c.50Hz to 15000Hz d.15000Hz to above

15. The pollution which does not persistent harm to life supporting system is ?

    a.Noise pollution   b.Radiation pollution

    c.Organochlorine pollution   d.All of these

16. Soap and detergents are the source of organic pollutants like ?

    a.Glycerol   b.Polyphosphates   c.All of these

17. Growing agricultural crops between rows of planted trees is known as ?

    a.Social forestry   b.Jhum   c.Taunga system   d.Agroforestry

18. The main atmospheric layer near the surface of earth is ?

    a.Troposphere   b.Mesphore   c.Ionosphere   d.Stratosphere

19. Name of substance whose accumulation in pelicians of lake Michigan led to

    the formation of thin shells of their eggs ?

    a.CFC   b.PAN   c.DDT   d.PAC

20. Name the process in which a harmful chemical enters the food chain and gets concentrated at each level in the food chain ?

    a.Concentration   b.Expansion   c.Pollution   d.Biomagnification

21. Specific enzymes are needed for the break down of a?

    a.Particular substance   b.Solid substance   d.None of these

22. Substances that are not break down in this manner are said to be ?

    a.biodegradable   b.Non biodegradable

    c.(a) and (b)   d.None of these

23. All organisms such as plants ,animals , microorganisms and human beings maintain a balance in ?

    a.Nature   b.Chain   c.None of these

24. Abiotic components comprising physical factors like ?

    a. Temperature   b.Rain fall   c.Soil   d.All of these

25. All living organisms interact with ?

    a.Each other   b.Together   c.Own   d.None of these

26. While gardens and crop fields are ?

    a. Human made ecosystem   b.Artifical ecosystem

    c. Both   d.None of these

27. The radiant energy of the sun in the presence of ?

    a. Chlorophyll   b.Photosythesis   c.None of these

28. All green plants and certain blue - green algae which can produce food by?

    a. Photosythesis   b.Sunlight   c.Water   d.None of these

29.   Organisms depend on the ?

a. Food   b.Plant   c.Producer   d.All of these

30.   Choose a correct statement ?

a. Tree – Goat – Tiger   b.Tree –Tiger – Lion

c. Tree – Elephant – Tiger   d.None of these

31.   Which are the first tropic level ?

a. Autotrophs , Hetrotrophs   b.Autotrophs , Producers

c. Plant and animal   d. None of these

32.   What is the example of harbivores ?

a. Goat   b.Lion   c.Tree   d. None of these

33.   What is the example of carnivores ?

a. Lion   b.Tiger   c.Man (human people)   d. All of these

34.   Man is the example of ?

a. Carnivores   b.Omnivores   c.Parasides   d.None of these

35.   Lice is the example of ?

a. Parasites   b.Omnivores   c.Carnivores   d.None of these

36.   In organic substances that go into the soil and are used up once more by the ?

a. Animal   b.Plant   c.(a)&(b)   d.None of these

37.   The food we eat acts as a fuel to provides us energy to do ?

a.Flight   b.Work   c.Sleep   d. All of these

38.   The autotrophs Capture the energy the energy present in sunlight convert it to?

a. Physical energy   b.Heat energy   c.Chemical energy   d. None of these

39.   Chemical energy Supports all the activities of the ?

a. Non living world   b.world living   c.Animal Plant   d.Plant

40. Some energy is lost to the environment in forms which can ?

    a. Be used again   b.Not be used again

    c.Not used energy   d.None of these

41. Green plants are eaten by ?

    a. Teritary consumers   b.Secondary consumers

    c. Primary consumers   d.Producers

42. Human beings occupy which level in any food chain ?

    a. Low level   b.Medium level   c.Top level   d.No level

43. Ozone is a molecule formed by three atoms of ?

    a.Carbon   b.Helium   c.Oxygen   d. Hydrogen

44. Ozone is a ?

    a.Deadly poision   b. Layer   c. Harmful   d.None of these

45. Which is highly damaging to organisms ?

    a.Ultravoilet radiation   b.Animal   c.Omnivores   d. Carnivores

46. Which is known to cause skin cancerin human beings ?

    a. Sunlight   b.Ultraviolet radiation   c.vitamin   d. None of these

47. The amount of ozone in the atmosphere began to drop sharply in the ?

    a.1980s   b.1990s   c.1970s   d.None of these

Answers:

Q	A	Q	A	Q	A	Q	A	Q	A
1	C	11	C	21	A	31	B	41	C
2	D	12	B	22	B	32	A	42	C
3	D	13	D	23	A	33	D	43	C
4	A	14	C	24	D	34	C	44	A
5	C	15	A	25	A	35	A	45	A
6	B	16	C	26	C	36	B	46	B
7	D	17	C	27	A	37	D	47	A
8	C	18	A	28	A	38	C		
9	B	19	C	29	C	39	B		
10	D	20	D	30	A	40	B		

10. NATURAL RESOURCES

SOME IMPORTANT POINTS

- Life exists on the earth because of resources like soil,water,air and the energy we get from the sun.
- The outer crust of the earth is called Lithosphere.
- Air is the mixture of many gases like nitrogen,oxygen,carbon dioxide and water vapors etc.
- Carbon dioxide is used by two ways;

  1. Green plants convert carbon dioxide into glucose in the presence of sunlight.

  2. Some marine animals use carbonates dissolve in sea water to make their shells.

- Uneven heating of air over land and water causes wind.
- Evaporation of water from water bodies and then condensation give us rain.
- Rainfall patterns depend on the patterns of wind.
- An increase of harmful substances in the air or environment is known as air pollution.
- The present source of fresh water is Rain.
- The improvement of harmful substances in the water is known as water pollution.
- The roots of the plants play an important role in preventing soil erosion.
- Nitrogen is an important compound for the synthesis of proteins,vitamins,DNA and RNA etc.
- The endoskeletons and exoskeletons of various animals are also formed from carbonate salts.
- Carbon-dioxide,Methane and Sulphur are known as the green house gases.

10. NATURAL RESOURCES

1.   The best requirements of all life forms on the earth are fulfilled by?

        (a) The resources available on earth

        (b) Energy from the sun

        (c) Atmosphere present on the earth

        (d) all of the above

2.   The outer crust on the earth is ?

        (a) Lithosphere   (b) stratosphere   c. biosphere   d. None

3.   What part of earth surface covered from water?

        (a) 25%   (b) 75%   c. 80%   d. None

4.   The biotic component of biosphere is?

        (a) living   (b) non living   c. both   d. None

5.   Living things are found where exists:

        a. atmosphere   b. hydrosphere   c. lithosphere d. None

6.   The air that covers the earth as a blanket is:

        a. hydrosphere b. atmosphere   c. biosphere   d. none of these

7.   The composition of air due to which life exist on earth is?

        a. oxygen   b. carbon dioxide   c. nitrogen   d. monoxide

8.   The cell who need oxygen to break down glucose molecule and

        get energy are?

        a. prokaryotic cell b. eukaryotic cell   c. both   d. none of these

9.   Life supporting of the earth where atmospheres hydrosphere and

        Lithosphere interact is?

        (a)biosphere   b. stratosphere   c. none of these   d. both a & b.

10.   The direction of wind during the day would be from?

        a. the sea of the land   b. the land of the sea

c. both   d. none of these

11. Winds are created due to the movement of?

a. air   b. water   c. both   d. none of these

12. Patterns of rain fall are decided by the prevailing of?

a. wind pattern   b. forest pattern   c. cloud pattern   d. both a & c

13. The mainly oxides which cause air pollution on earth are?

a. sodium   b. sulphur   c. nitrogen   d. both b & c

14. When the harmful gases dissolve in rain its cause?

a. acid rain   b. tsunami   c. both   d. none

15. The visible indication of air pollution is?

a. smog   b. acid rain   c. both   d. none

16. Fresh water are available in form of?

(a)ice –caps   (b)underground water

(c)both (a)and (b)   (d)none of these

17. What factors causes water pollution on the earth?

(a)dissolving of fertilizers and pesticides in water

(b) sewage from our towns

(c) all of the above

18. The water that is found inside the deep reservoir would be?

(a)cold   (b)hot (c)cold than upper part (d)none of these

19. Which factors helps in making of soil?

(a) the sun   (b)water   (c)wind (d)all of above

20. Some bits of decayed living organisms are also found in the soil are?

(a)humus (b)minerals (c)nutrients (d)none of these

21. Which factor determines that which plants are grown on soil?

(a) Nutrients of soil   (b) humus present in soil

(c) depth of the soil   (d)all of these

22.   The topmost lawyer of soil is?

(a)top soil   (b)medium soil (c)none of these (d)both (a)and (b)

23.   Which organism helps in making rich humus?

(a)earthworms (b)bacteria (c)fungi (d)none of these

24.   Which factors are responsible for the removal of fine particles from soil?

(a)flowing of water   (b)wind (c)both (a)and (b) (c)none of these

25.   The nitrogen fixing bacteria are most commonly used in the roots of?

(a) Legumes   (b) carrot (c) photo python (d) none of these

26.   Nitrates and nitrites are converted into amino acids to make?

(a)proteins (b)carbohydrates (c)fats (d)none of these

27.   The combined state of carbon is?

(a) Carbon dioxide   (b) carbonates

(c) Hydrogen –carbonate salts   (d)all of these

28.   The green house gases which lead to global warning are?

(a) Carbon dioxide   (b) methane (c) both (a) and (b) (d) oxygen

29.   The oxygen from the atmosphere can be used by?

(a) Respiration (b) combustion (c) formation of oxides (d) all of above

30.   The formulae of ozone is?

(a)O_2 (b) O   (c) O3   (d) none of these

31.   A hole in the ozone layer was first seen above?

(a)Antarctica   (b) ammonia (c) South Africa   (d) none of these

32.   Ozone layer gets depleted by the used of?

(a)CFCS   (b) hydrogen gas   (c) carbon dioxide gas (d) none of these

33.   The ozone layer protects us from?

(a)U.V rays   (b) harmful gases   (c) tsunami   (d) none of these

34.   Nitrogen is contained in?

(a)nucleic acids (b)amino acids (c)fats (d)none of these

35.   In elemental form carbon access as?

(a)diamonds   (b)graphite   (c)both (a)and (b) (c)none of these

ANSWERS:

QUES.	ANS.	QUES.	ANS.	QUES.	ANS.	QUES.	ANS.
1	d	11	a	21	D	31	A
2	a	12	d	22	A	32	A
3	b	13	d	23	A	33	A
4	a	14	a	24	C	34	A
5	d	15	C	25	A	35	C
6	b	16	C	26	A		
7	a	17	D	27	D		
8	C	18	C	28	C		
9	A	19	D	29	D		
10	a	20	a	30	C		

11. MANAGEMENT OF NATURAL RESOURCES

SOME IMPORTANT POINTS

- ➤ We need to manage our natural resources by sustainable methods.
- ➤ The demand for all resources is increasing at a expanded rate.
- ➤ The management of resources helps not to be exploited and live last for long time.
- ➤ There are many things we can do to manage our natural resources out of them three R s major practice. (Reduce, Reuse, Recycle).
- ➤ The sustainable method to protect forest resources should be taken in the favour of stakeholders.
- ➤ Stakeholders are those people who live near the forest and who are totally dependent on forest resources.
- ➤ We need to use the fossil fuels coal and petroleum in very large rare cases, because they will ultimately exhaust.
- ➤ Watershed management not only increases the production and income of the watershed community but also mitigates droughts and floods and increases the life of downstream dams and reservoirs.
- ➤ We need to reduce our requirements, so that the benefits of development reach everyone for now and for all generations to cone.

11. MANAGEMENT OF NATURAL RESOURCES

1. Which one of the following is an example of biotic component of environment?

   (a)Wind   (b)Water   (c)Vegetation   (d)Temperature

2. Which of the following is an non renewable resource?

   (a)Solar energy   (b)Hydrocarbon fuel   (c)Flora and Fauna (d)Nuclear

3. Sanctuaries are established to?

   (a)Conduct ecotourism on wildlife   (b)Protect animals (c)None of these

4. Global warming has resulted due to ?

   (a)Lack of rainfall wlorldwide (b)Increased emissions of CO_2 form automobiles

   (c)Oxides of sulphur and nitrogen   (d)None of these

5. The main source of water in India is?

   (a)Rain water   (b) Ground water   (c)Surface water   (d)Sea water

6. Floods are caused by ?

   (a)Afforestation   (b)Cutting the forests

   (c)Tilling the land   (d)None of these

7. The ganga runs from gangotri through a hundred towns and cities ?

   (a)U.P   (b)U.P and Bihar

   (c)U.P ,Bihar,W.B   (d)U.P,W.B,Haryana

8. Water pollution can be identified by testing its ?

   (a)PH level   (b)Biological oxygen demand

   (c)Both   (d)None of these

9. The three R'S to save the environment are ?

(a)Reserve,Reduce ,Recycle (b)Reuse ,Reserve,Reduce

(c)Reserve,Reuse,Reduce   (d)Reduce,Rcycle ,

10.   Which is in Rajasthan?

(a)Surangams   (b)kattas   (c) Kulks   (d)Nadis

11.   Kattas is the ancient method of water harvesting found in ?

(a)Himzachal pardesh   (b)Karnatka

(c)Tamil nadu   (d)M.P

12.   Amirata devi bishnoi sacrificed her life to the protection of the ?

(a)Sal trees   (b)Pine trees

(c) Khejri trees   (d)Alpine meadows

13.   In independent india Plantation of which trees caused their monocultures ?

(a)Eucalyptus   (b) Pine

(c)Eucalyptus ,Pine and teak (d)Eucalyptus, Pine, teak and Need

14.   The Ganga runs its course from ?

(a)Ganga sagar   (b)Himalays Peak Everest

(c)Gangotri   (d)Jamnutri

15.   The presence of which micro organisms in ganga water indicates

Contamination ?

(a)Amoeba   (b)Mucar spores

(c)Coliform bacteria   (d)None of these

16.   The Chipko movement started from ?

(a)Reni in gharwal   (b)Arborio forest   (c)Khejrali village

17.   Primary source of water is ?

(a)Rivers   (b)Ground water   (c)Lakes   (d)Rain water

18.   Tawa irrigation project is in ?

(a)Maharastra   (b)Mahdhya pardesh   (c)Orissa   (d)Haryana

19.   Measure of biodiversity of an area is ?

(a)The number of species found there (b)The range of different life forms

(c)Both (a)&(b)   (d)None of these

20.   Which energy of water is used to produce hydro electricity ?

(a)Potential energy   (b) kinetic energy (c) Both (a)&(b)   (d)None of these

21.   Chipko Andolan is concerned with ?

(a)Conservation of natural resourses

 (b)Zoological survey of India

(c)Forest conservation

22.   The concept of 'Biosphere Reserves' was evolved by?

(a)Government of India   (b)Botanical survey of India

(c)UNESCO   (d)UNDP

23.   Why should be conserve biodiversity because?

(a)We should preserve biodiversity we have inherited

(b)A loos of diversity may lead to aloss of ecological stability

(c)Both(a)&(b)

24.   The problems for critism about large dams are that they?

(a)Displace large number of peasants and trebles without proper rehabitation

(b)Contribute enormously to deforestation and the loss the biological diversity

(c)All of the above

25. Natural resourses like?

(a)Soil   (b)Air   (c)Water   (d)All of these

26. Multi-crore project came about in?

(a)1986   (b)1985   (c)1988   (d)1990

27. The Ganga runs its course of over?

(a)2600 Km   (b)2700 Km   (c)2500 Km   (d)2575 km

28. Largely untreated sewageis dumped into the Ganga every?

(a)Day   (b)Month   (c)Year   (d)All of the days

29. The pollutants are?

(a)Useful   (b)Harmful   (c)Both   (d)None of these

30. Will ultimately be exhausted?

(a)Fossil fuels   (b)light   (c)Petroleum   (d) (a)&(c)

31. Petroleum is used?

(a)Car & Bike   (b) Cycle, lamp   (c) None of these

32. Coal and petroleum have been formed from?

(a)Gobar gas   (b)Biomass   (c)None of these

33. Water harvesting is an age – old concept in?

(a)India   (b)Kerala   (c)America   (d)China

34. Parts of Himachal Pradesh had evolved local system of canal irrigation called?

(a)Kulhs   (b)Ponds   (c) Surangams   (d)Eris

35. kulhs is used for?

(a)Irrigation   (b)Drink   (c)None of these

36.   Rains in India are laregely due to the?

(a)Monsoons   (b)Winter   (c)Wind   (d) None of these

37.   If the goals of all the stakeholders with regards to the management of the forest are?

(a)Medium   (b)Same   (c)High   (d)None of these

38.   What is used to make slats for huts?

(a)Plants   (b)Wood   (c)Bamboo   (d)None of these

39.   Coal is used in?

(a)Thermal power (b)Chemical idustury (c)Physical laboratory (d)None of these

40.   Who was started Chipko Andolan?

(a)Ramlal Bahuguna   (b) Sunder lal Bahuguna

(c)Ramesh lal Bahuguna   (d)Shyam lal Bahuguna.

Answers :

Q	A	Q	A	Q	A	Q	A	Q	A	Q	A	Q	A	Q	A
1	C	6	C	11	B	16	A	21	A	26	B	31	A	36	A
2	B	7	C	12	C	17	C	22	C	27	C	32	B	37	B
3	B	8	C	13	C	18	B	23	C	28	D	33	A	38	C
4	B	9	D	14	C	19	C	24	C	29	B	34	A	39	A
5	A	10	D	15	C	20	A	25	D	30	D	35	A	40	B

SOURCE OF ENERGY

SOME IMPORTANT POINTS

➢ Energy neither be created or destroy its only courted from one form to another
➢ A good source of energy would be that have high calorific value be easily accessible be easy to store and transport be eco-friendly
➢ Energy are characterized into two types
  1. Conventional source of energy
  2. Non-conventional source of energy
➢ Fossil fuels, thermal power plant, hydro-power plant, bio-mass and wind energy are the examples of conventional source of energy
➢ Solar energy tidal energy, wave energy, ocean thermal energy, nuclear energy, and geo-thermal energy are the examples of non-conventional source of energy
➢ Coal petroleum and natural gas the example of now-renewable source of energy
➢ Air, water, solar ration and tidal energy are the example of renewable source of energy
➢ hydro power- plants converts the potential energy of thee falling water into electricity by making dams
➢ animals waster plant waste and other bio degradable wastes used for making bio-mass
➢ bio-gas is a excellent fuels contains 75% of methane
➢ to is stalling a wind mill we required a very large area and the minimum speed of wind is 20km/h
➢ solar cells made up of silicon and its converts light energy into electrical energy
➢ nuclear energy is divided into two types
  1. Nuclear fission
  2. Nuclear fusion
➢ the splitting of heavy atom into many lighter atoms to gives energy is known as nuclear fission

- two of more atoms combined to makes a very atom to gives energy
- the energy stores as heat inside the earth is used to rotate the turbine and then to the generator converts heat energy into electrical energy is known as geo-thermal energy

12. SOURCES OF ENERGY

1. The chemical energy in the wax converted to
    heat energy and light energy on?
    (a) burning   (b) washing
    (c) cooling   (d) none of these

2. Sun's energy is due to nuclear?
    (a) fission   (b) fusion   (c) collaps   (d) none of these

3. wood was the comman source of ?
    (a) chemical energy   (b) heat energy
    (c) electrical energy   (d) muscular energy

4. The growing demand for energy was largely met by the fossil fuels?
    (a) coal   (b) petroleum
    (c) a & b   (d) none of these

5. Coal and petroleum are?
    (a) limited   (b) unlimited   (c) none of these

6. The fossil fuels are which sources of energy?
    (a) Renewable   (b) non renewable   (c) none of these

7. Burning fossil fuels has other?
    (a) Disadvantages too   (b) advantages too
    (c) non advantages   (d) none of these

8. The air pollution caused by burning of which products?
    (a) Coal   (b) petroleum
    (c) both   (d) none of these

9. ------------- rain which effects our water and soil resources?
    (a) acid   (b) base
    (c) neutral   (d) none of these

10. Fossil fuels are the major fuels used for?
    (a) Heat energy   (b) chemical energy
    (c) generating electricity   (d) none of these

11. Charcoal burns without?
    (a) heat   (b) flames   (c) none of these

12. The starting material is mainly cow –dung, it is known as?
    (a) bio –dung   (b) gobar gas   (c) a & b   (d) none of these

13. Wind energy is an environment friendly and efficient source of?

(a) renewable energy   (b) heat energy   (c) non renewable energy   (d) none of these

14. The wind speed should also be higher than?

    (a) 15kn/h   (b) 10km/h   (c) 20km/h   (d) 30km/h

15. For a 1Mw generator the farm needs about?

    (a) 2 hectares of land   (b) 3 hectares of land

    (c) 4 hectares of land   (d) none of these

16. The initial cost of establishment of the farm is quite?

    (a) technological   (b) electrical

    (c) mechanical   (d) none of these

17. Which progress our demand for energy increases day by day?

    (a) technical   (b) electrical

    (c) mechanical   (d) none of these

18. Which we use to do more and more of our tasks?

    (a) Machines   (b) hard work

    (c) hand   (d) all of these

19. Our demand for energy?

    (a) increases   (b) decreases

    (c) remains constant   (d) none of these

20. India receives the energy equivalent to more than?

    (a) 5000 million kwh   (b) 5000 billion kwh

    (c) 5000 trillion kwh   (d) none of these

21. The average distance between the sun and earth is called?

    (a) Energy constant   (b) heat constant

    (c) Solar constant   (d) light constant

22. A large number of solar cells are combined in an arrangement of, it is known as?

    (a) solar cell panel   (b) solar system

    (c) battery   (c) none of these

23. Solar energy is?

    (a) limited   (b) unlimited

    (c) constant   (d) none of these

24. What is the example of solar – cell panel?

    (a) TV   (b) radio

    (c) none of these

25. The sea – shore can be trapped in a manner to generate?

    (a) Electricity energy   (b) heat

    (c) Energy   (d) none of these

26. Renewable energy is available in our?

    (a) Natural environment   (b) science laboratory

    (c) none of these

27. A solar water heater can be used to get hot water on?

    (a) A sunny day   (b) a cloudy day

    (c) A wind day   (d) all day

28. Many of the sources ultimately derive their energy from the?

    (a) Solar   (b) bio- mass

    (c) Moon   (d) sun

29. Green house effect is caused by gases like?

    (a) helium   (b) oxygen   (c) nitrogen   (d) CO_2

30. A solar water heater cannot be used to get hot water on?

    (a) A sunny day   (b) a cloudy day

    (c) A hot day   (d) a windy day

31. Chief constituent of natural gas is?

(a) Methane (b) ethane (c) butane (d) propane

32. The power plant which converts potential energy of falling

   Water into electricity is?

   (a) Nuclear plant   (b) thermal plant

   (c) Hydro plant   (d) wind plant

33. Wood is a

   (a) Primary fuel   (b) secondary fuel

   (c) Liquid fuel   (d) processed fuel

34. Which of the following is not an example of a bio-mass

   Energy source?

   (a) wood   (b) gobar gas   (c) coal   (d) nuclear

35. The popular name of biogas is?

   (a) gobar gas   (b) marsh gas

   (c) ethane gas   (d) helium gas

36. The country of winds is?

   (a) India   (b) china   (c) Denmark   (d) Netherlands

37. Which of the following is the ultimate source of energy?

   (a) water   (b) uranium   (c) sun   (d) fossils fuel

38. Acid rain happens because?

   (a) sun leads to heating of upper layer of atmosphere

   (b) Burning of fossil fuels release oxides of carbon nitrogen

   and sulphur in the atmosphere

   (c) Earth atmosphere contains acids

(d) None of these

39. Which one of the following forms of energy leads to least environmental pollution on the process of its harneshing and utization?

(a) nuclear energy   (b) thermal energy

(c) Solar energy   (d) geothermal energy

40. The major problem in harneshing nuclear energy is how to   (a) split nuclei   (b) sustain the reaction

(c) dispose off spent fuel safely (d) none of these

41. The major (gas) constituent of biogas is?

(a) Methane   (b) carbon dioxide

(c) Hydrogen   (d) hydrogen Sulphide

42. Which part of the solar cooker is responsible for green house effect?

(a) mirror (b) glass   (c) black colour (d) none of these

43. Fuel used in thermal power plant is?

(a) water (b) uranium   (c) biomass (d) fossil fuels

44. Large eco systems are destroyed when submerged under the water?

(a) In dams   (b) on dams

(c) Over the dam   (d) none of these

45. Opposition to the construction of Tehri dam on which river are due to problems?

(a) Yamuna (b) Narmada (c) ganga (d) Brahmaputra

46. Fuels are of which product?

(a) plant (b) animal   (c) plant & animal (d) none of these

47. Plant has which like structure built with bricks?
(a) dome   (b) dame   (c) dam   (d) none of these

48. The digestor is a sealed chamber in which there is no?

(a) oxygen   (b) carbon   (c) helium   (d) methane

49. Bio gas is an excellent fuel as it contains up to ----------% methane?

(a) 73%   (b) 75%   (c) 80%   (d) 85%

50. Methane heating capacity is?

(a) Low   (b) high   (c) none of these

51. Bio gas used for?

(a) Lighting (b) plant   (c) animal (d) none of these

52. Bio mass is a which source of energy?

(a) Renewable (b) non- renewable (c) heat energy

(d) Chemical energy

53. In a water lifting pump the rotatory motion of windmill is utililsed to lift water from a well is an example of?

(a) Kinetic energy   (b) energy

(c) Heat energy   (d) none of these

54 wind energy is also used to?

(a) Generate electricity   (b) electrical energy

(c) heat energy   (d)chemical energy

55. A number of windmills are arranged over a large area is

Called

(a) Wind energy   (b) heat energy

(c) Chemical energy   (d) none of these

Answers:

Q.	A.	Q.	A.	Q.	A.	Q.	A.	Q.	A.	Q.	A.
1	A	11	B	21	C	31	A	41	A	51	A
2	B	12	B	22	A	32	C	42	B	52	A
3	B	13	A	23	B	33	A	43	D	53	A
4	C	14	A	24	C	34	D	44	A	54	A
5	A	15	A	25	A	35	A	45	C	55	A
6	B	16	B	26	A	36	C	46	C		
7	A	17	A	27	A	37	C	47	A		
8	C	18	A	28	D	38	B	48	A		
9	A	19	A	29	D	39	C	49	B		
10	C	20	C	30	B	40	C	50	B		

13. IMPROVEMENTS IN FOOD NUTRIENTS

SOME IMPORTANT POINTS

- Plants too require nutrients to fulfill their growth demands. There are thirteen nutrients which are essential for plants and which are supplied from soil to the plants.
- Macro - nutrients out of thirteen nutrients, six nutrients are required by plants in large quantities, which are called macro – nutrients.
- Micro – nutrients remaining seven nutrients are required in small quantities which are called micro nutrients.
- Manure and fertilizers are one of the main sources of nutrients, supplied to plants.
- The organic farming is a type of farming in which organic manures, recycled farm wastes and bio – agents with healthy cropping systems are used.
- Farming of two types of crops together on a same piece of land is called mixed cropping.
- The growing of two or more types of crops in a row pattern is known as Inter – cropping.
- The farming of different crops on a piece of land in a planned successive pattern is known as crop rotation.
- Animal husbandry is type of complete care for farm animals, such as shelter, feeding, breeding and disease.
- Fishes are obtained from both marine and inland resources.
- The echo sounds and satellites are used to guide the fishing nets to capture marine fish.
- The composite fish culture system is a best place for fish farming.
- The capacity of honey collection is very in Italian bees.
- The Indian bees that are used for the production of honey are apiscerana India.
- Poultry farming is undertaken to raise domestic fowl for egg production and chicken, meat.

IMPROVENENT IN FOOD RESOURCES

1. The berseem crop is a type of :-

    (a) rabi season crop   (b) fodder crop

    (c) kharif season crop   (d) none of these

2. Wheat are fall in which category of crop:-

    (a) rabi season crop   (b) fodder crop

    (c) kharif season crop   (d) none of these

3. The period of kharif season crop is starts from:-

    (a) November to july   (b) june to October

    (c) march to December   (d) none of these

4. The crossing between different species to same genus refers to:-

    (a) pasturage   (b) wider adaptability

    (c) Hybridisation   (d) none of these

5. Which type of nutrients plants not get from the soil:

    (a) hydrogen and chlorine

     (b) carbon and oxygen

    (c) nitrogen and phosphorus

    (d) none of these

6. Which category of nutrients related to zinc.

    (a) micro-nutrients   (b) manure nutrients

    (c) macro-nutrients   (d) none of these

7. Which pair of crops not used in mixed cropping :-

(a) wheat-groundnut   (b) pulse-mustard

(c) wheat-maize   (d) none of these

8. Which cattle breeds are used for increasing lactation period.

(a) red sindhi + sahiwal

(b) jerrey + brown swiss

(c) brown swiss + sajowal

(d) none of these

9. Which fishes are commonly called as bottom feeder:-

(a) mrigals   (b) rohu

(c) starfish   (d) lobster

10. Which breed of bee is used for commercial honey production:-

(a) A. dorsata   (b) A. florae

(c) A. meligera   (d) none of these

11. Which of the following are fresh water fish:-

(a) rohu and mrigals
(b) starfishand lobster

(c) rohu and dphins

(d) none of these

12. Which is net falls in category of cereal crop

(a) wheat   (b) mustard

(c) maize   (d) none of these

13. Pigeon pea is generally known as:-

(a) chana

(b) moong

(c) arhar

(d) masoor

14.	Which in not falls in category of oil sed crops:-

(a) soyabean

(b) lentil

(c) castor

(d) none of these

15.	Which crop is not used as a food for the live stoclc:-

(a) linseed

(b) oats

(c) sudan gass

(d) none of these

16.	Which is not falls in the category of khasib crops

(a) maize

(b) paddy

(c) wheat

(d) none of these

17.	Which crop shocus the characteristics of tallness and profuse braching.

(a) oil

(b) fodder

(c) cereal

(d) none of these

18.	Assel is known as a desi breed of

(a) cow

(b) sheep

(c) poultry

(d) none of these

19.	Which breed of bee is known as rock bee:-

(a) apis florae

(b) apis dorsata

(c) apis serena

(d) none of these

20.	The method of bish farming is also known as:-

(a) culture fishery

(b) fish adaptation

(c) capture fishing

(d) none of these

21. Which of the following is a finned fish:-

    (a) mullets   (b) mollusks

    (c) starfish   (d) none of these

22. The capacity of honey collecting are very high in:-

    (a) Indian bees   (b) italian bees

    (c) german bees   (d) none of these

23. Which are of the following is Indian breed

    (a) assel   (b)   leghorn

    (c) cerana   (d) none of these

24. Which one of the following is foreign breed of cock

    (a) assel   (b) meghorn

    (c) cerana   (d) none of these

25. Which is the point where sea water and fresh water mix together

    (a) brakish water   (b) salt water

    (c) prawn water   (d) none of these

Q.	A.	Q.	A.	Q.	A.	Q.	A.	Q.	A.
1	B	6	A	11	A	16	C	21	A
2	A	7	C	12	B	17	B	22	B
3	B	8	B	13	C	18	C	23	A
4	C	9	A	14	B	19	B	24	B
5	B	10	C	15	A	20	A	25	A

NOTES